AF292197

ENERGY

By
Robin Twiddy

BookLife
PUBLISHING

©2023
BookLife Publishing Ltd.
King's Lynn, Norfolk
PE30 4LS, UK

A catalogue record for this book is available from the British Library.

ISBN: 978-1-80155-748-1
ISBN: 978-1-80155-854-9

Written by:
Robin Twiddy

Edited by:
Madeline Tyler

Designed by:
Amy Li

All facts, statistics, web addresses and URLs in this book were verified as valid and accurate at time of writing. No responsibility for any changes to external websites or references can be accepted by either the author or publisher.

IMAGE CREDITS

CONTENTS

Words that look like this can be found in the glossary on page 24.

WHAT IS ENERGY?

Energy is something that makes other things work. Energy powers lots of different things.

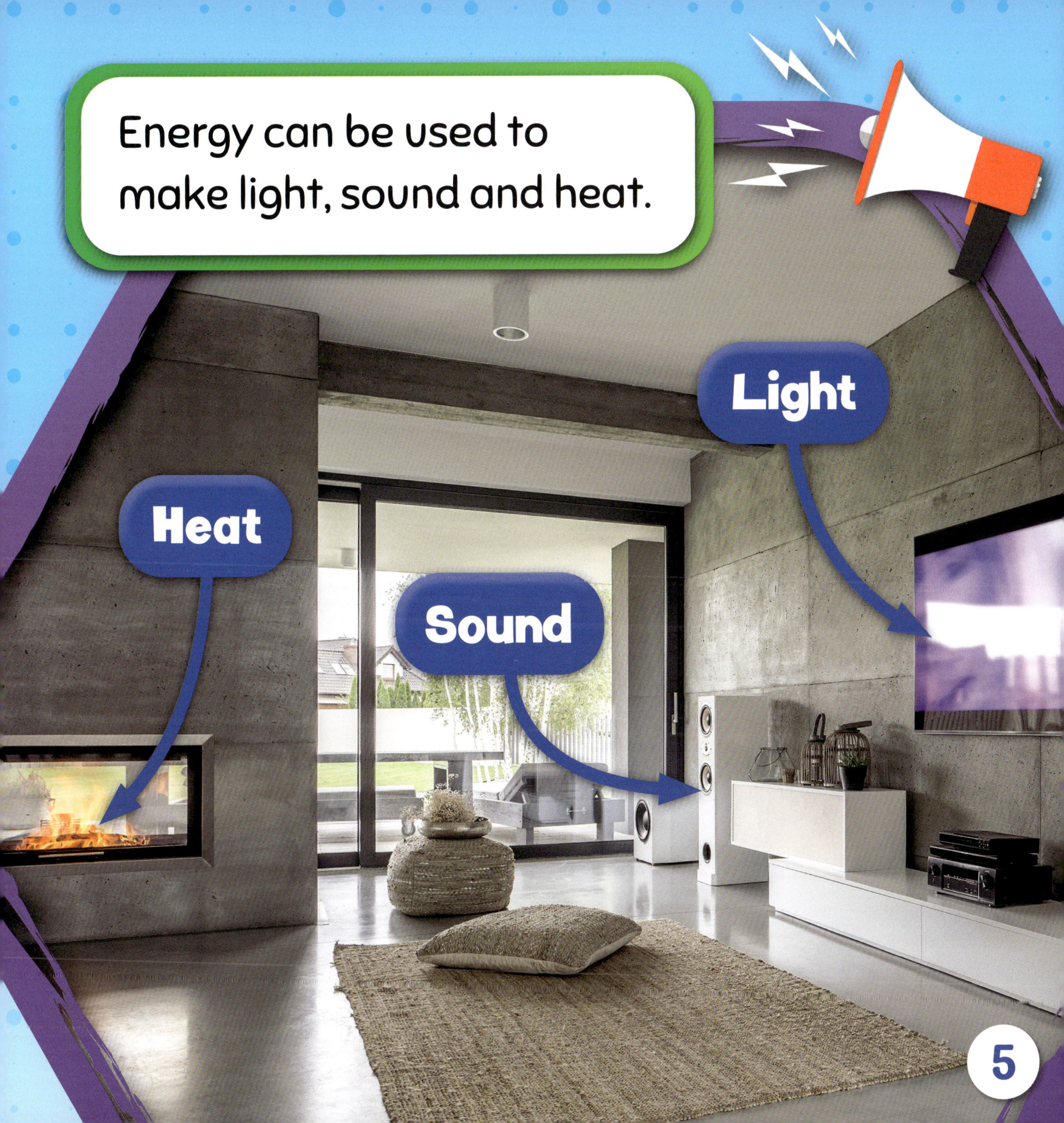

Energy can be used to make light, sound and heat.
Light
Heat
Sound
5

DIFFERENT TYPES OF ENERGY

Electricity is one type of energy, but there are many more.

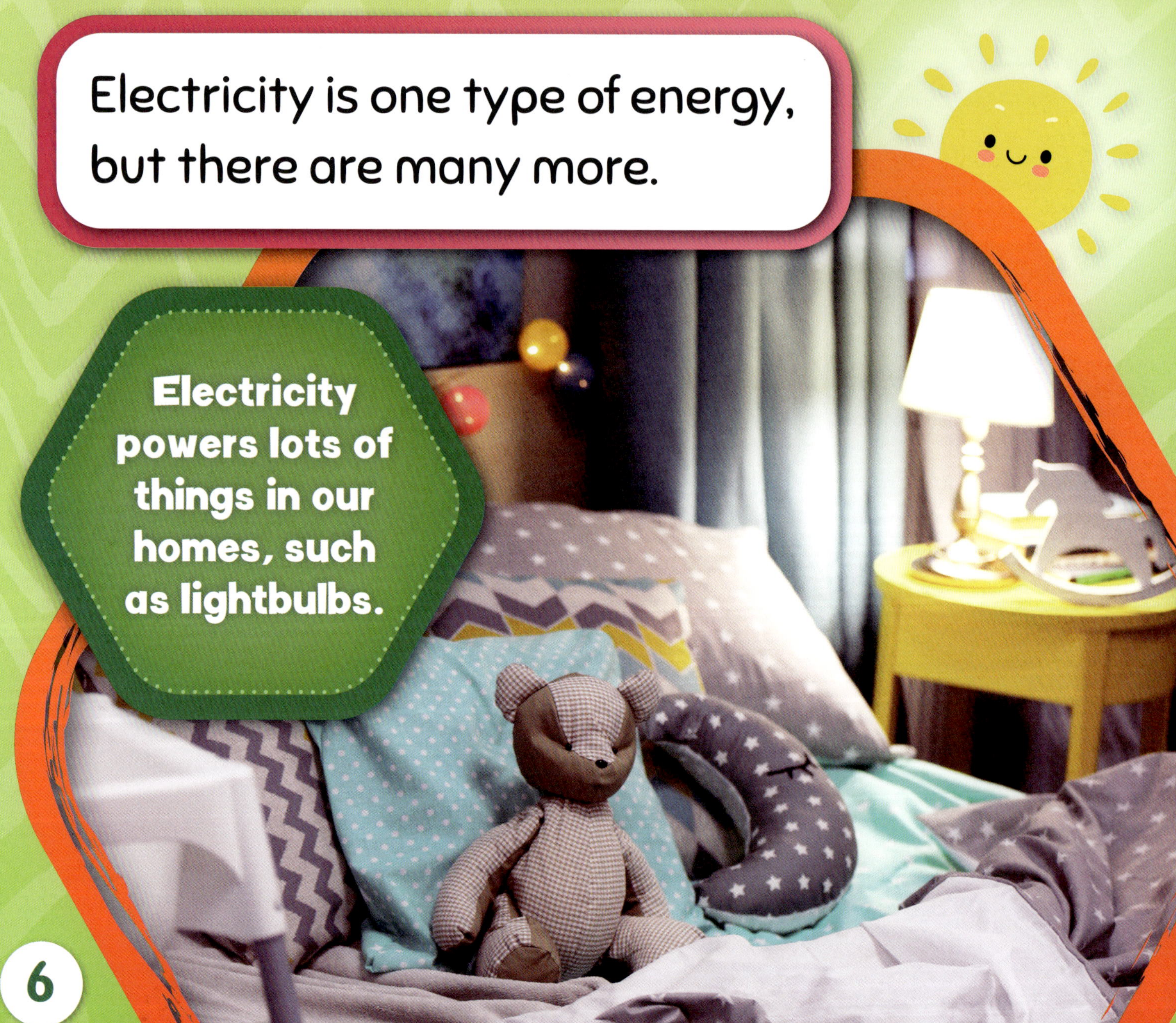

Anything that moves or grows uses energy to do it.
This plant uses energy to grow.
7

ELECTRICITY

Electricity is a type of energy. There are things powered by electricity all around your home.

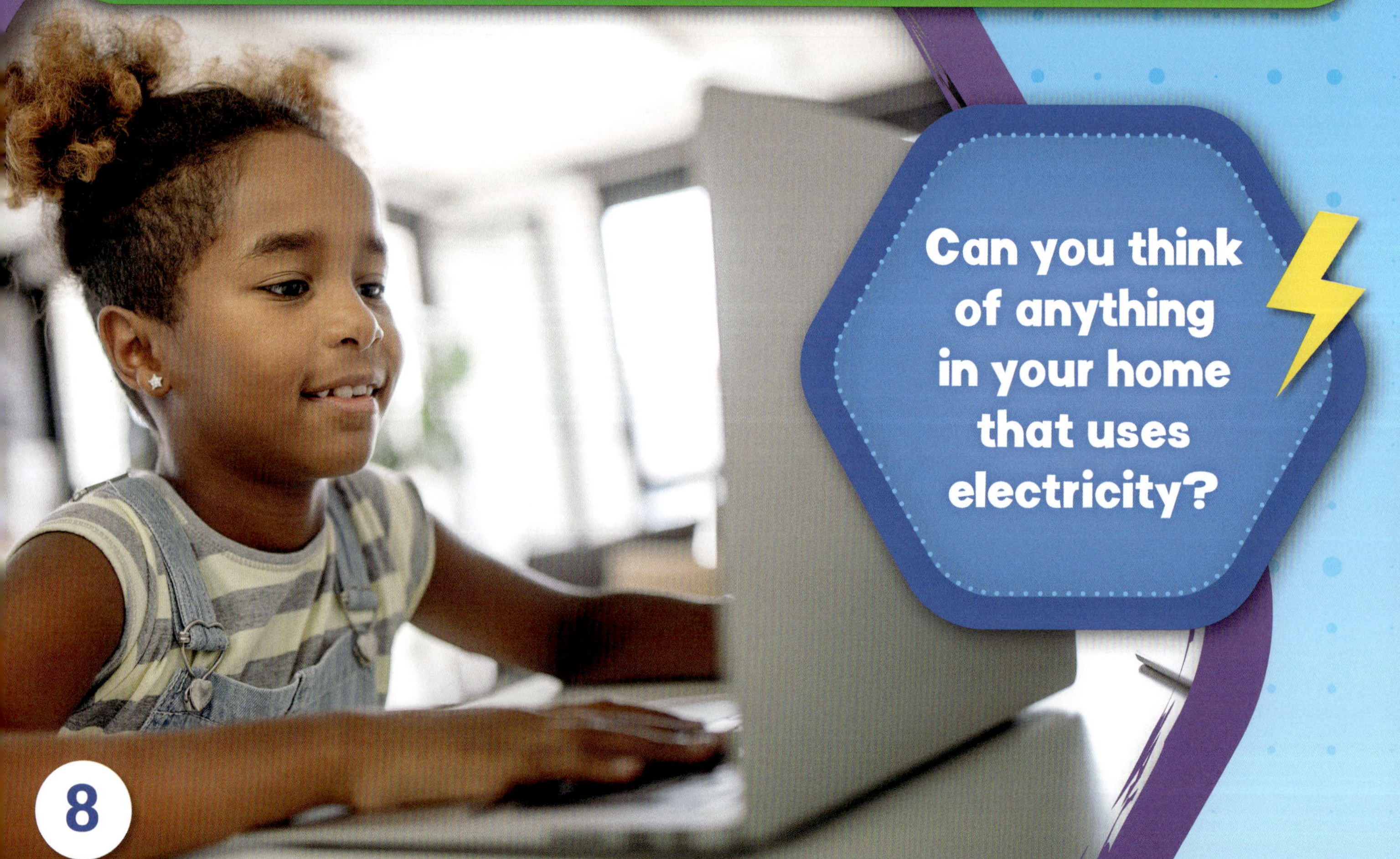

Electrical items can be powered by batteries or by being plugged into the wall.
Batteries
Electrical socket
Always ask an adult to plug things in for you.

HOW IS ENERGY MADE?

Energy can be made by using things from the ground, such as coal, or by using wind, water or sunlight.

Some things that we use to make energy will run out after a while. Other things can be used over and over.

FOSSIL FUELS

Coal, oil and gas are all <u>fossil fuels</u>. Electricity is made by burning these in power stations.

Fossil fuels are not renewable.

Burning fossil fuels is not good for our planet because it creates <u>pollution</u>.

NUCLEAR

Energy is sometimes made in nuclear power stations.

Nuclear power stations are very large. They can cause big problems when they go wrong.

RENEWABLE ENERGY

SOLAR

We can use the energy from the Sun to make power. This is known as solar power.

Solar energy will not run out!

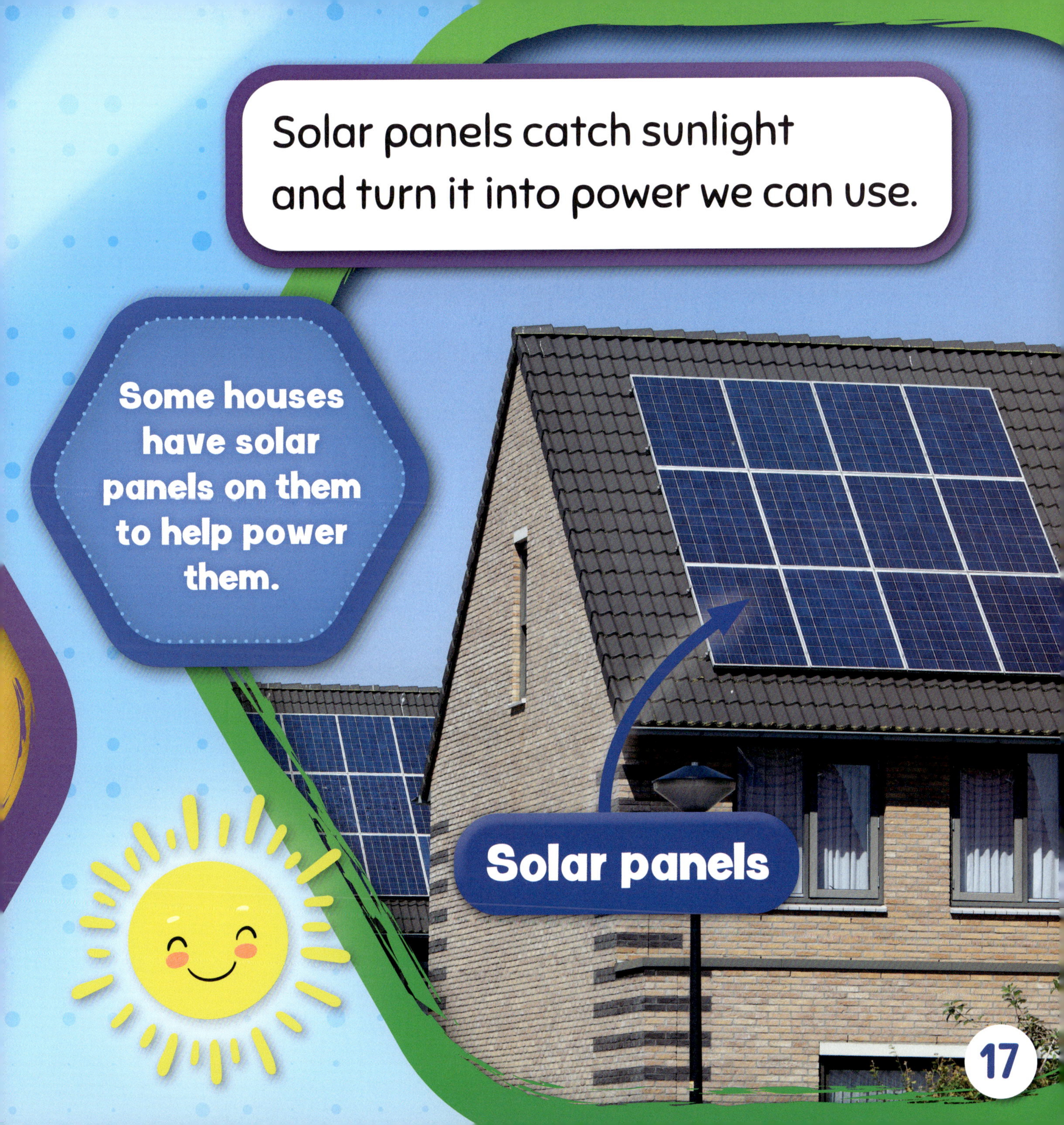

Solar panels catch sunlight and turn it into power we can use.
Some houses have solar panels on them to help power them.
Solar panels

17

HYDROPOWER

We can make energy from running water. This is known as hydropower.

Water is a way of making energy that will not run out.

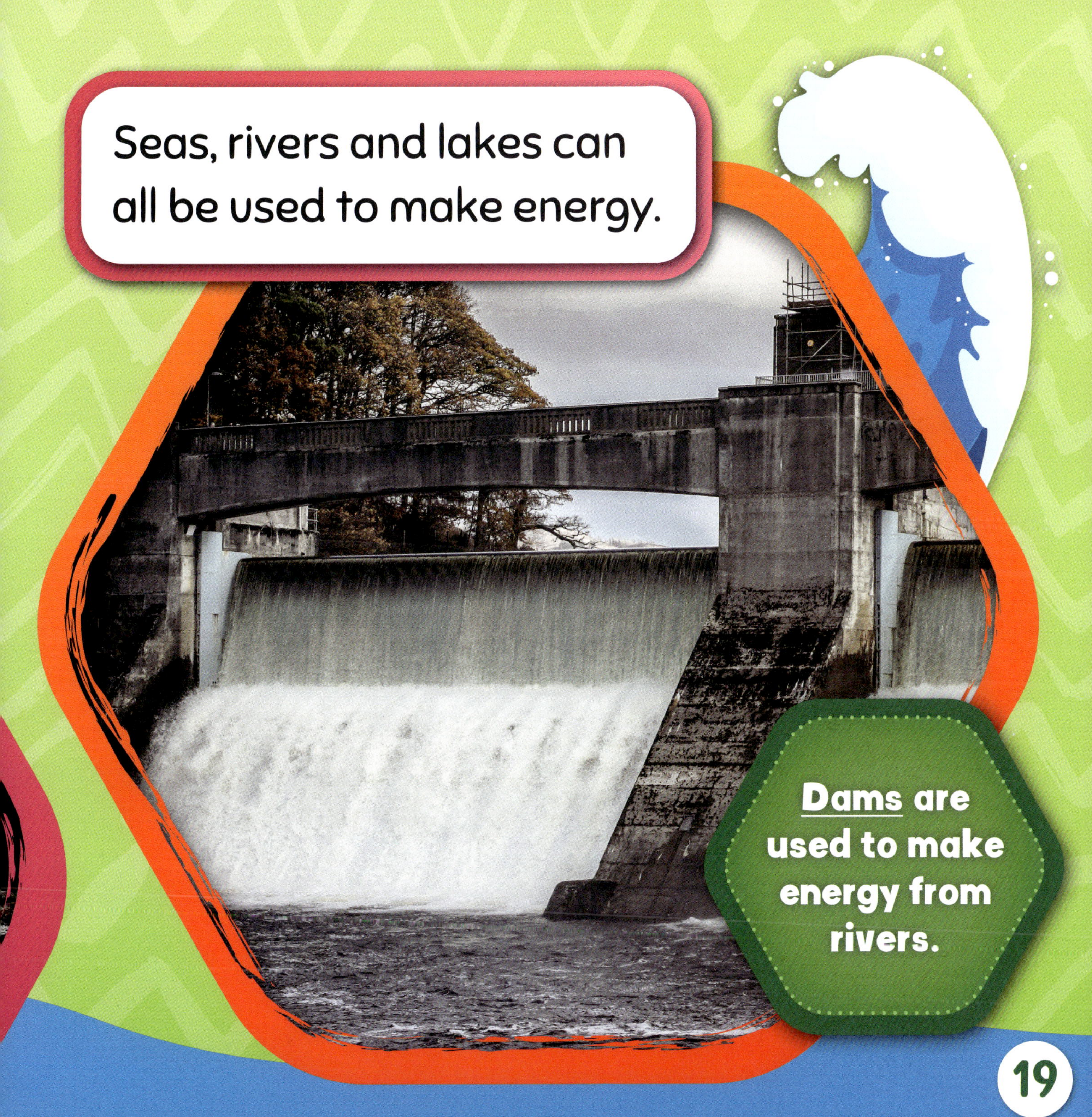

Seas, rivers and lakes can all be used to make energy.

Dams are used to make energy from rivers.

19

WIND POWER

Energy can be made from wind by using wind <u>turbines</u>.

The wind pushes the blades of the turbine around.

The energy used by the wind to spin the blades is turned into electricity.

ENERGY AND OUR PLANET

Some of the ways that we make energy can hurt our planet.

If we look after our planet, it will look after us, too.

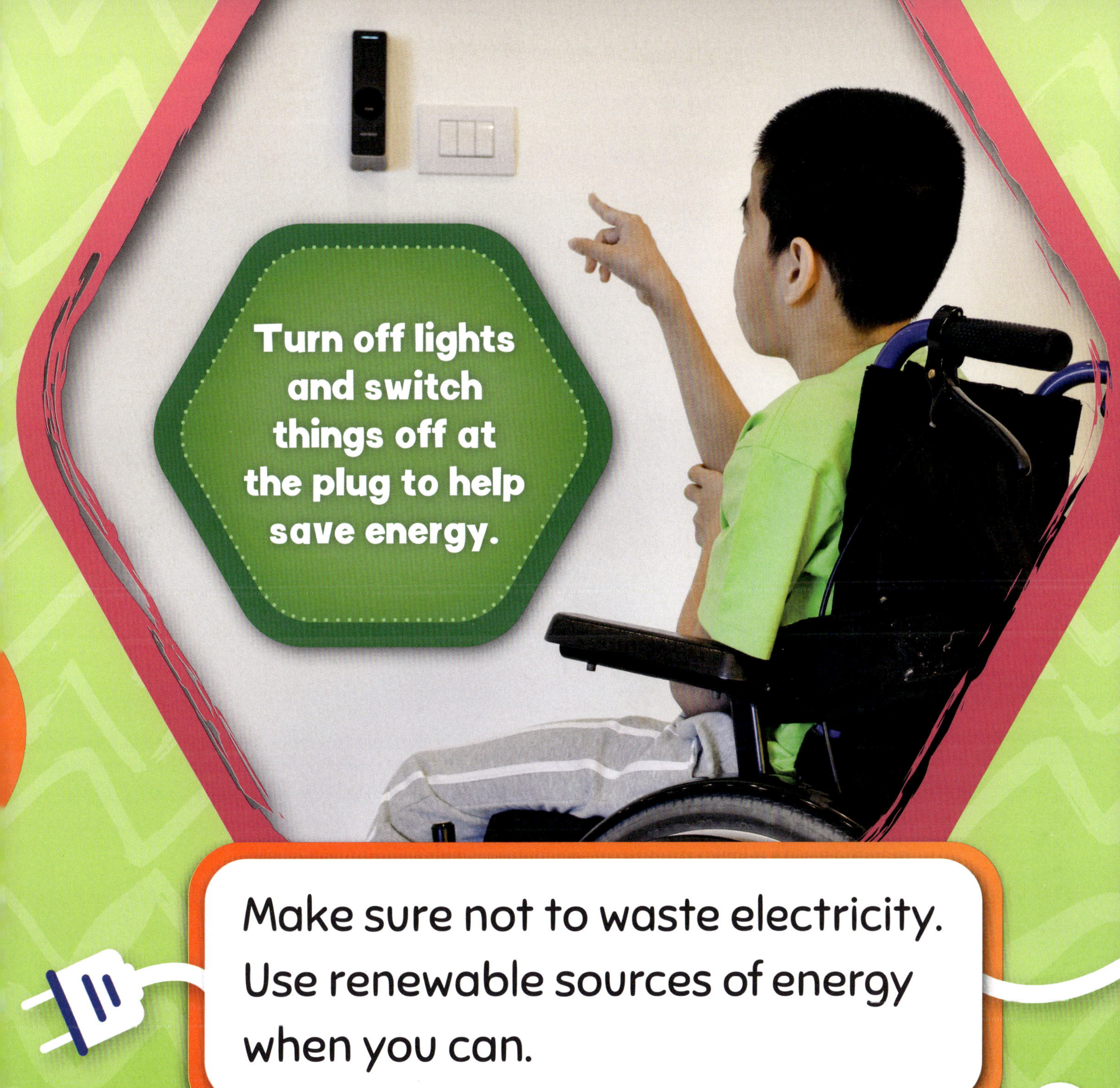

Make sure not to waste electricity. Use renewable sources of energy when you can.

GLOSSARY

DAMS	large walls that block rivers and streams
FOSSIL FUELS	fossilised materials that are millions of years old and have large amounts of energy stored in them
POLLUTION	when harmful and poisonous things are added to an environment
RENEWABLE	able to be used again without it running out
TURBINES	machines moved by water, steam or air that make power

INDEX